Annie Eluo

Numerical prediction of rainfall over the city of Kisangani

Annie Eluo

Numerical prediction of rainfall over the city of Kisangani

Study conducted in the communes of Kabondo and Kisangani

ScienciaScripts

Imprint

Any brand names and product names mentioned in this book are subject to trademark, brand or patent protection and are trademarks or registered trademarks of their respective holders. The use of brand names, product names, common names, trade names, product descriptions etc. even without a particular marking in this work is in no way to be construed to mean that such names may be regarded as unrestricted in respect of trademark and brand protection legislation and could thus be used by anyone.

Cover image: www.ingimage.com

This book is a translation from the original published under ISBN 978-620-2-28163-8.

Publisher:
Sciencia Scripts
is a trademark of
Dodo Books Indian Ocean Ltd. and OmniScriptum S.R.L publishing group

120 High Road, East Finchley, London, N2 9ED, United Kingdom
Str. Armeneasca 28/1, office 1, Chisinau MD-2012, Republic of Moldova, Europe
Printed at: see last page
ISBN: 978-620-5-87633-6

Table of contents

SPATIAL MODELLING OF RAINFALL IN THE CITY OF KISANGANI, CASE OF THE COMMUNES OF KABONDO AND KISANGANI.

SUMMARY

Within the framework of the adaptation to the climatic changes, by advocating a study of remote sensing of the city, to identify and measure the permanently floodable and flooded zones by associating them with the climatic models of the pluviometric predictions it was possible to register on our maps processed by GIS under Arc Gis the curves of the levels SRTM, and to understand the leveling of the floods below the threshold of 400m of altitude on the city of Kisangani.

1. INTRODUCTION

The city of Kisangani is largely uncontrolled urbanization. It is also characterized by very intense rainfall. In addition, the soils are made up of a thick sandy alteration mantle. The conjunction of all these factors makes this city the seat of spectacular erosion of floods. Thus, the problem of soil loss by gullying is extremely acute (LUBUIMI, 2003). The presence of disasters caused by the occupation of sites at risk: "Floods, silting, pollution, erosion and landslides" and the lack of resources for better urban management are frequent (ANONYMOUS, 2006).

Indeed, the ponds, rivers and shallows of the communes of Kabondo and Kisangani do not escape this sad situation.

reality. As a result of overpopulation, all marginal lands are occupied, including hazardous areas and erosion-prone areas.

"The deforestation in the region of Kisangani has now had negative consequences on the city, with its corollary of danger such as floods.

At this stage of knowledge, this work proposes the following solutions to bet on the inevitable disaster that are on the social level:

- to unclog the sewers of the city of Kisangani,

- to fight against the anarchic constructions on the city,

- To educate the population of the Eastern province and in particular that of Kisangani so that it becomes aware of the importance of the flora. All this, to fight against deforestation. In addition, in rural areas, people must also begin to plant trees to replace the woods that they are cutting in a disorderly way.

In addition, from a scientific point of view, it is appropriate to think about 3D contour mapping, to follow the watershed of rainwater, as well as the geo-positioning of the ponds in the city to be coupled with monthly rainfall data. Such a study will inform us on the trajectory of water following the contours, the speed of the water currents of the rains as well as the filling of the ponds below the DTM.

This is, in our humble opinion, what needs to be done to fight against flooding and at the same time against deforestation because the two problems are linked.

Is there a scientific explanation for the Kisangani flood? We think so, because in the era of climate change and recurrent variability of extreme weather events, Africans must fight not only to <u>mitigate</u> the negative effects of climate in their immediate environment, but especially to <u>adapt to it</u>. Thus, in the world, the mitigation of severe climate must go through reforestation, the fight against deforestation, increased use of renewable and clean energy, waste treatment, rehabilitation of gutters and water pipes in cities. In addition, adaptation to climate change requires, among other things, the construction of dykes and water retention basins, the cultivation of wood energy, the fight against poverty by converting ponds into fish and rice growing areas, wastewater treatment, the prevention of waterborne diseases and the financing of studies on climate forecasts and modelling in areas at risk such as Kisangani

2. MATERIAL AND METHOD

The materials used for this work are the following:

2.1.1. Satellite images

We used the following images:

A.   SRTM (contour) images of Kisangani ;

B.   Landsat images of the city of Kisangani.

These satellite images (SRTM and LANDSAT) were processed at the scale of ponds, streams and shallows with appropriate mapping software (ARC GIS 10, QGIS2.0).

All these images were obtained at the OSFAC Laboratory and at the Department of Soil Physics and Hydrology of the Commissariat Général à l'Energie Atomique in Kinshasa through C.T. Jean Fiston MIKWA and Prof Albert KABASELE.

A 30 m resolution SRTM image was used to generate the floodplain contours and stream DTM.

> Arc GIS 10 has allowed us to publish dynamic maps more efficiently and to facilitate the sharing of geographic information.

> QGIS 2.0 allowed us to implement the GPS data and to produce shapes and polygons of the communes of the city of Kisangani and the water ponds as well as to digitize the locations of these water ponds.

2.1.2. Rainfall data

We used the data collected at the meteorology division for the rainfall of Kisangani from 1956 to 2008 and the monthly rainfall of these two communes to make the curves in histograms and evaluate the periods at risk of flooding.

2.1.3. Field materials

We collected data on the ground with a GPS by taking points (Longitude, Latitude and Altitude) on the boundaries of the communes, the water ponds, the shallows as well as the watercourses in the communes of Kabondo and Kisangani. We also used motorcycle cabs for transportation.

2.2 METHODS

The methodology used is the following and can be summarized in 5 main steps: 1).field data collection; 2).image acquisition; 3).rainfall data processing; 4).software processing; 5).data processing.

The documentary approach allowed us to learn about the concept of spatial modeling, the Geographic Information Systems approach to flood risk studies, the remote sensing approach to flood monitoring, the Water Ponds approach as well as the integration of rainfall data parameters and field data into a GIS.

2.2.1. Field data collection

The data collection was done through field visits to the two communes (Kisangani and Kabondo) from January 13 to February 5, 2014; This data collection was done by taking the geographical data of each water pond (stagnant water, puddle, fishponds, swamp) located in the two communes as well as the rivers and shallows using a GPS Garmin Type 62S, Accuracy 3m for the In Situ Classification while waiting for the appearance of at least four satellites; but the first trip was to identify the precise locations of the ponds in the two communes by inquiring in the offices of the communes and the second was to collect the data.

To identify the precise locations of the ponds, rivers, and shallows, we had to visit the administrative offices of these two communes, specifically the rural development and civil status departments, to find out the boundaries of these two communes; we also visited the office of the Ministry of Planning and the Geographic Institute to see if they had maps of the city of Kisangani and of these two communes, but unfortunately we did not find any.

The field survey was used to delineate the major environmental problems identified in the field and to establish their relationship with the geometric, morphometric and hydrological characteristics assessed with the software.

We then went to the office of the provincial division of meteorology to collect climatic data of the city of Kisangani from 1956 to 2005 in order to build an umbrothermal curve with these data and to calculate the volume of water in these Ponds.

2.2.2. Image acquisition.

A 30 m resolution SRTM image was used to generate flood contours and stream DTMs.

Acquisition of rainfall data for the city of Kisangani.

We went to the meteorological service to take the data of the rainfall of Kisangani from 1956 to 2008 and the monthly rainfall (precipitation) of these two communes to make the diagrams and evaluate the zones at risk of flooding including the months with extreme rainfall. With these data we have

• Daily rainfall;

determined :

- monthly rainfall ;

- annual rainfall ;

- the average annual rainfall modulus (arithmetic mean of the annual rainfall amounts) ;

- the monthly rainfall fraction (ratio between the annual module and the monthly module considered) ;

- averages, average number of rainy days, variability of rainfall and rainy days. With all these data we will calculate the average monthly rainfall of Kisangani.

- Monthly and annual rainfall maps

- Volumes of water in water ponds.

- Monthly forecasting equations.

With : Monthly precipitation and Annual precipitation.

The maximum monthly rainfall was recorded in months with extreme rainfall such as November and the driest month such as January.

2.2.3. Software Processing

We used the following software:

- With Arc GIS 9.3, we have increased our overall GIS productivity.

Arc GIS 10 integrates tools and functionalities to help us in :

- efficient management of spatial information,

- the creation of professional quality cards,

- the sharing of operational information,

- Anthropization (ecosystem deforestation, wetland drainage),

- Natural evolution of the environment,

- Real-time information exchange, from and to the field,

- planning and analysis.

Arc GIS 9.3 has allowed us to publish dynamic maps more efficiently and to facilitate the sharing of Geographic Information.

> QGIS allowed us to implement the GPS data and produce shapes and polygons of the communes and ponds as well as to identify the locations of the ponds from the field data by digitizing them.

2.2.4. Data processing

We have processed the data collected on the field with the help of GPS by downloading it on the machine, we have also processed the rainfall data and the GIS data but first we have had to pre-process these SRTM images in order to correct them; for the SRTM to eliminate the negative values of altitude (attributable to the taking of the data by the satellite)

Mastery of software manipulation to facilitate the processing of these data.

Finally, all these data were integrated into a GIS to evaluate the flood risk areas and the months with extreme rainfall in these two municipalities and to make umbrothermal curves.

The data processing was done by computer and on the internet.

3. RESULTS AND DISCUSSION

111.1. GEO-LOCATION OF WATER PONDS IN THE COMMUNES OF KABONDO AND KISANGANI IN THE CITY OF KISANGANI.

Figure 1 and 2. The water ponds listed in the Commune Kabondo; Scale: 1:15392 Commune Kisangani and its water ponds; Scale: 1:12535.

These two figures show that there are permanently flooded areas in the Kabondo Commune, represented by a cluster of ponds (swamps, shallows, stagnant water...) with a total area of 41 km.) with a total surface of $41Km^2$ on the $368km^2$ of the total surface of the commune and this agglomeration of water ponds was digitized on QGIS according to their locations (3 agglomerations: that of the Tshuapa District; that of the Kibibi District and that fed by the mukoko river) and georeferenced on ArcMap, each of which is represented on the following maps. We have indicated there beside its location in relation to the Democratic Republic of Congo and the city of Kisangani. The commune has a surface of 12m2 and a density of 15521, the surface is of 368 Km2. And in the commune Kisangani the places where the ponds of waters were located were repertoried with the help of GPS after the descent on ground and it was noticed that the aforementioned Commune comprises zones permanently floodable with the presence of an agglomeration of ponds of waters of approximately six places (quarttier kongakonga I and II,avenue walikale I and II, an agglomeration on 10th and 11th avenues, on 13th avenue mosibasiba district) with a surface of 17,62km2 approximately 120km2 on 636 km2 of the totality of the Commune.Let us point out that the commune kisangani has a density of 12298m2 with 7m2 of the total surface.

111.2. PRESENTATION OF THE WATER LEVEL CURVES OF THE KABONDO AND KISANGANI COMMUNES

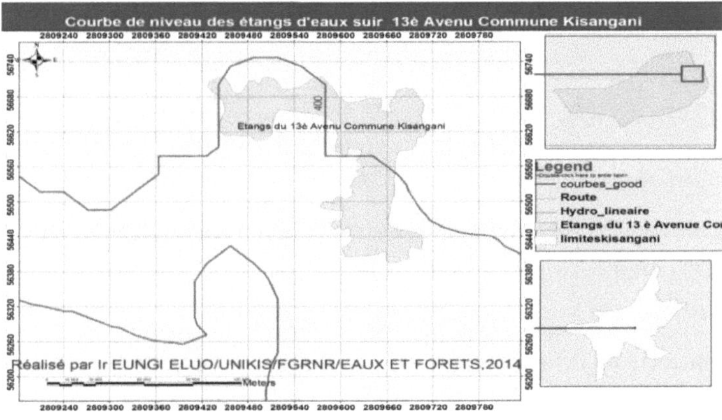

Figure 3. The water level curve of the ponds on 13ème avenue mosibasiba district

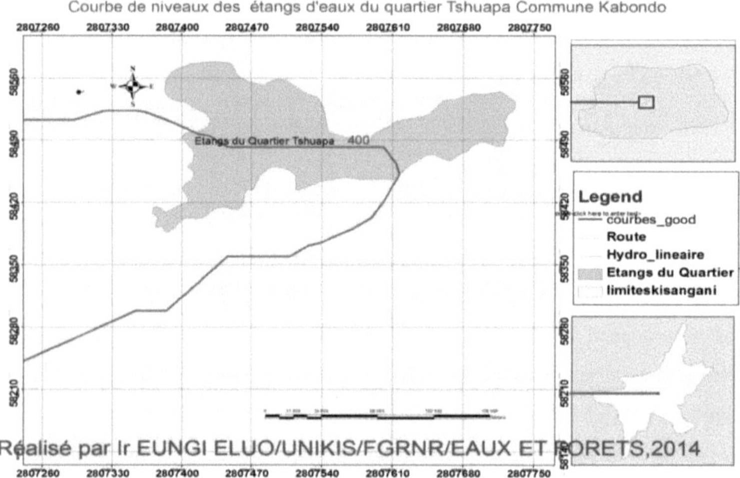

Figure 4. The water level curve of the ponds in the Tshuapa district of Kabondo Commune.

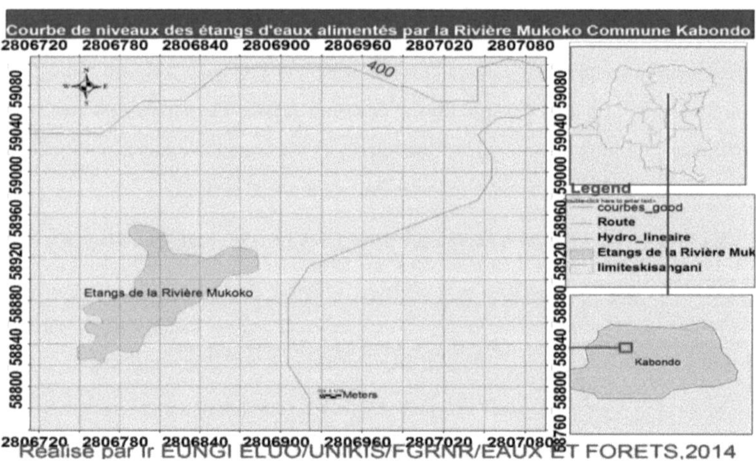

Figure 5. The water level curve of the ponds fed by the Mukoko River, Kisangani Commune.

It appears from all these contour lines that the water ponds of these two municipalities are all in an altitude lower than 400m; what at about 375m as the minimum of the lowest level. This is the basis for the flooding of the city.

112. . PRESENTATION OF MONTHLY RAINFALL DIAGRAMS FROM 1956 TO 2008.

Descriptive Statistics of Monthly Precipitation Kisangani City in 52 years Source Eungi lr2Unikis 2014

	N	Minimum	Maximum	Mean	Std. Deviation
Januaries	52	13.80	182.60	81.0442	45.541 83
FevPluies	52	20.80	196.30	97.1000	42.15289
MarchRains	52	46.30	299.30	1.4106E2	56.10866
AprilRains	52	60.50	315.10	1.8282E2	58.4081 1
MayRains	52	81.10	311.00	1.6433E2	53.061 03
JuneRains	52	39.60	302.00	1.1852E2	53.60688
JullPluies	52	46.70	234.00	1.1003E2	38.8221 3
AugustRains	52	44.40	286.00	1.4058E2	48.80039
Seven Rainfalls	52	80.30	337.50	1.8099E2	58.63398
OctRains	52	94.50	375.50	2.2264E2	59.3551 1
NovPluies	52	34.20	388.30	1.9479E2	66.8691 5
DecPluies	52	29.40	346.20	1.2898E2	65.42602
Valid N (listwise)	52				

Table 1: Descriptive statistics of monthly prescriptions in the city of Kisangani in 52 years.

Comment:

1.　N = 52 years of rainfall data from Kisangani 1956 -2008

2.　Minimum, Maximum and Mean in mmH2O = LitreH20/m2 from January to December over 52 years

3.　Jan =January, Feb = February, Jull= July, Sept = September, Oct = October, Nov= November, Dec= December and, Valid N Listwise = Total validated N

4.　Std.Deviation = Standard Deviation

5.　Monthly rainfall = Mean ± Standard deviation

6.　Minimum = Lowest monthly value for 52 years

7.　Maximum= Highest monthly value for 52 years

8.　Mean = 1.4106E2 = 1.4106x100 =141.06 for example

DISCUSSION OF THE RESULTS

a). Monthly rainfall retention capacity.

Water ponds of the Tshuapa district Commune Kabondo

Means in liter/m^2 ; Daily volume=Area x Means

Month	Area (m2) per GIS Remote sensing	Monthly rainfall (liter/m2)	Monthly volume (liter)

January	28459	81,0442	2306436,89
February	28459	97,1000	2763368 ,9
March	28459	141,06	4014426 ,54
April	28459	182 ,82	5202874 ,38
May	28459	164,33	4676667,47
June	28459	118,52	3372960,68
July	28459	110,03	3131343,77
August	28459	140,58	4000766,22
September	28459	180,99	5150794,41
October	28459	222,64	6336111,76
November	28459	194 ,79	5543528,61
December	28459	128,98	3670641,82
Total			46.155.495,1liter/year

Tableau 2   The volume of rainwater in the ponds of the Tshuapa district of Kabondo Commune.

The table shows that the water ponds in the Tshuapa district contain 46,155,495.1 liters of water per year

EauETANG

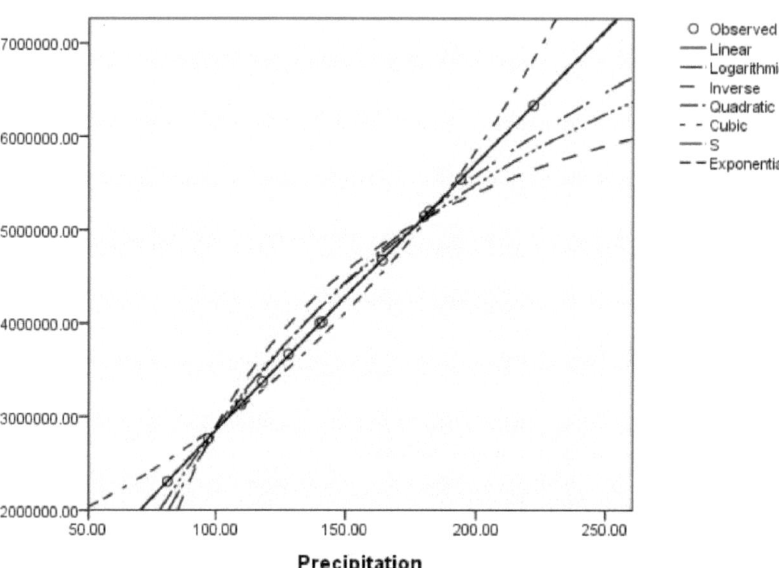

Figure 6: Graph of the relationship between the volume of water and the amount of rainfall on the ponds in the Tshuapa district of Kabondo.

Month	Area (m^2) by GIS Remote sensing	Rainfall monthly (liter/m2)	Monthly volume (liter)
January	7174	81,0442	581411,091
February	7174	97,1000	696595,4
March	7174	141,06	1011964,44
April	7174	182 ,82	1311550,68
May	7174	164,33	1178903,42
June	7174	118,52	850262,48
July	7174	110,03	789355,22
August	7174	140,58	1008520,92
September	7174	180,99	1298422,26
October	7174	222,64	1597219,36
November	7174	194 ,79	1397423,46
December	7174	128,98	925302,52
Total			12.646.931,2litres/an

Tableau 3   *Volume of water in the ponds of the Kibibi district.*

This table shows us that the water ponds in the Kibibi district of Kabondo Commune contain 12646931.2 liters of water/year of its volume.

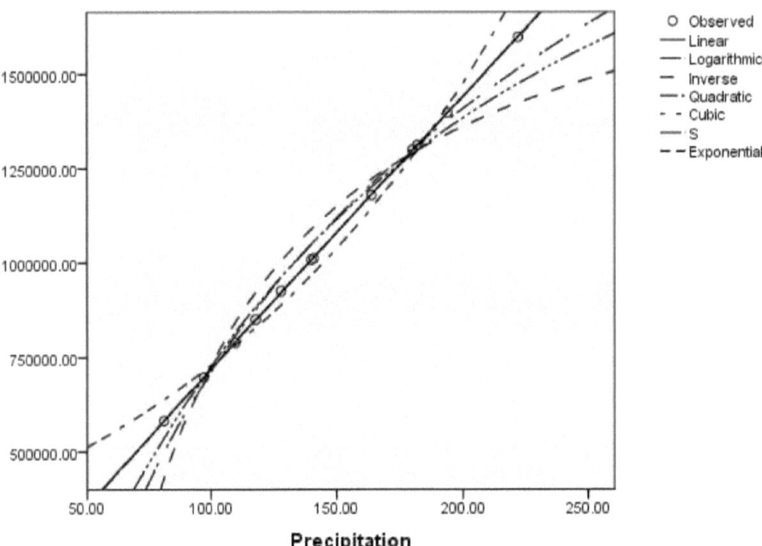

Figure 7: Graph showing the relationship between the volume of water and the amount of rainfall on the water ponds in the Kibibi district of Kabondo commune.

4.2.1.3. Water ponds fed by . the Mukoko River Commune Kabondo

Month	Area (m^2) per GIS Remote sensing	Monthly rainfall (liter/m2)	Monthly volume (liter)
January	6476	81,0442	524842,2392
February	6476	97,1000	6288819,6
March	6476	141,06	913504,56
April	6476	182 ,82	1183942,32
May	6476	164,33	1064201,08
June	6476	118,52	767535,52
July	6476	110,03	712554,28
August	6476	140,58	910396,08
September	6476	180,99	1172091,24
October	6476	222,64	1441816,64
November	6476	194 ,79	1261460,04
December	6476	128,98	835274,48
Total			17076438.08litres/year

Tableau 4 *volume of water from the ponds fed by the Mukoko River.*

The volume of water found in these ponds is 17,076,438.08 liters of water per year.

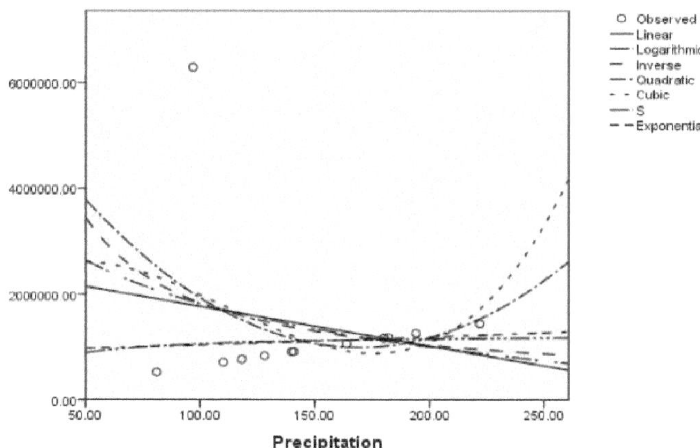

Figure 8: Graph of the relationship between the volume of water and the amount of rainfall on the ponds fed by the Mukoko River.

4.2.1.4. The water ponds on 13ème Avenue/ Mosibasiba district Kisangani Commune

Month	Area (m^2) by GIS Remote sensing	Rainfall monthly (liter/m2)	Monthly volume (liter)
January	33142	81,0442	2685966,876
February	33142	97,1000	3218088,2
March	33142	141,06	4675010,52
April	33142	182 ,82	6059020,44
May	33142	164,33	5446224,86
June	33142	118,52	3927989,84
July	33142	110,03	3646614,26
August	33142	140,58	4659102,36
September	33142	180,99	5998370,58
October	33142	222,64	7378734,88
November	33142	194 ,79	6455730,18
December	33142	128,98	4274655,16
Total			49838415,96

Tableau 5 Volume of water in the water ponds of the 13th avenue of Kisangani.

It regales that these water ponds contain 49.838.415,96 liters of water per year on a surface of $33142m^2$.

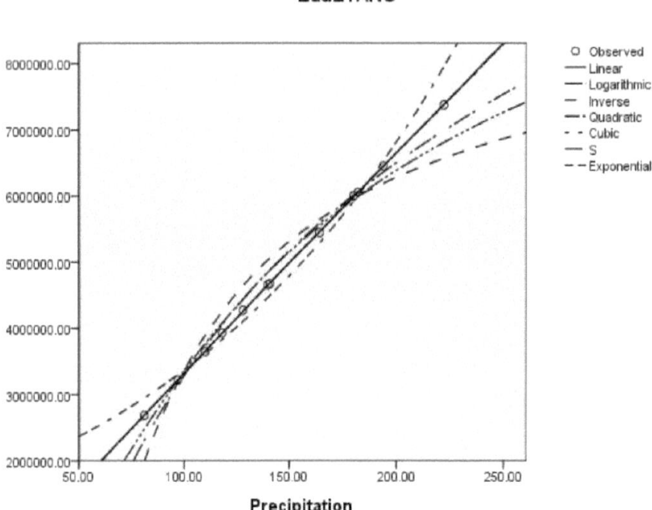

Figure 9. This graph explains the relationship between the amount of rainfall (rainfall data) and the volume of water in the ponds at 13ème avenue/quartier mosibasiba commune Kisangani (remote sensing).

4.2.1.5. The water ponds on 10ème and 11ème avenues in Kisangani commune.

Month	Area (m^2) per GIS Remote sensing	Monthly rainfall (liter/m2)	Monthly volume (liter)
January	44709	81,0442	3623405,138
February	44709	97,1000	4341243,9
March	44709	141,06	6306651,54
April	44709	182 ,82	8173699,38
May	44709	164,33	7347029,97
June	44709	118,52	5298910,68
July	44709	110,03	4919331,27
August	44709	140,58	6285191,22
September	44709	180,99	8091881,91
October	44709	222,64	9954011,76
November	44709	194 ,79	8708866,11
December	44709	128,98	8708866,11

Total			73.050.222,88litres/an

Table 6: Volume of water in the ponds of the 10th and 11th Avenues, Kisangani communes

It regales that these ponds contain 73.050.222,88litres of water per year.

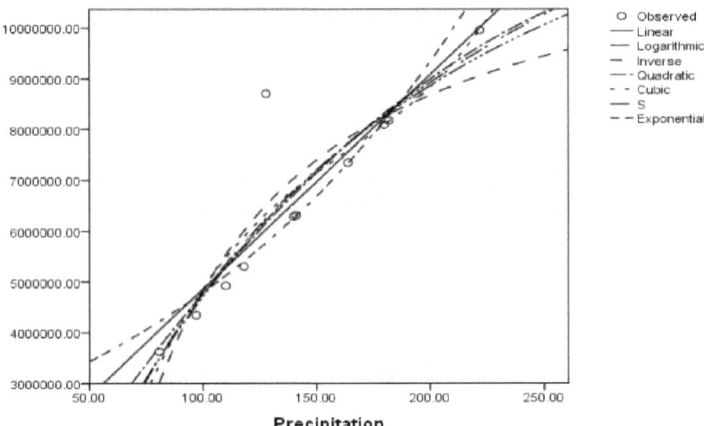

Figure 10. This graph explains the relationship between the amount of rainfall (rainfall data) and the volume of water in the water ponds on $10^{\text{ème}}$ and $11^{\text{ème}}$ avenues, commune Kisangani (remote sensing).

4.2.1.6. The water ponds in the Kongakonga district of Kisangani Commune

Month	Area (m^2) per GIS Remote sensing	Rainfall monthly (liter/m2)	Volume monthly (liter)
January	7915	81,0442	641464,843
February	7915	97,1000	768546,5
March	7915	141,06	1116489,9
April	7915	182 ,82	1447020,3
May	7915	164,33	1300671,95
June	7915	118,52	938085,8
July	7915	110,03	870887,45
August	7915	140,58	1112690,7
September	7915	180,99	1432535,85
October	7915	222,64	1762195,6
November	7915	194 ,79	1541762,85
December	7915	128,98	1020876,7

Total			10758496,99

Tableau 7 Volume of water from the ponds on 10ème and 11ème avenues, Kisangani commune.

The volume of water in these ponds is 10,758,496.99 liters of water on an area of 7915m^2 .

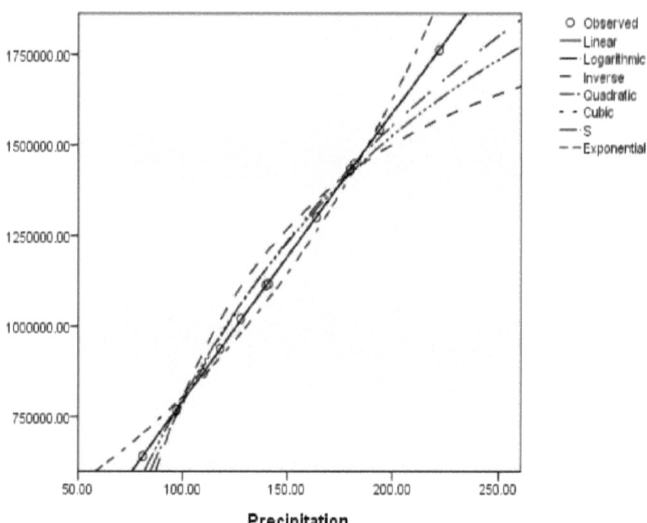

Figure 11. This graph explains the relationship between the amount of rainfall (rainfall data) and the volume of water in the ponds of Kongakonga, commune Kisangani (remote sensing).

4.1.2.7. The water ponds in the KongakongaI district of Kisangani Commune (see figure 15).

Month	Area (m^2) per GIS Remote sensing	Rainfall monthly (liter/m2)	Volume monthly (liter)
January	20832	81,0442	1688312,774
February	20832	97,1000	2022787,2
March	20832	141,06	2938561,92
April	20832	182 ,82	3808506,24
May	20832	164,33	3423322,56
June	20832	118,52	2469008,64
July	20832	110,03	2292144,96
August	20832	140,58	2928562,56
September	20832	180,99	3770383,68
October	20832	222,64	4638036,48

November	20832	194 ,79	4057865,28
December	20832	128,98	2686911,36
Total			33785842,41

Tableau 8 Volume of water.

The table above shows the volume of water in the ponds of the Kongakonga district which is 33,785,842.41 liters of water per year with a surface area of 20832m2.

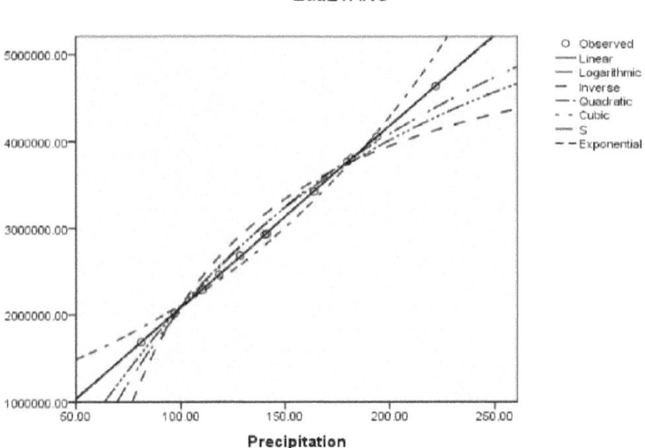

Figure 12. This graph explains the relationship between the amount of rainfall (rainfall data) and the volume of water in the ponds in Kongakonga II (remote sensing)

5.1.2.8. The water ponds on Walikale Avenue in Kisangani

Month	Area (m^2) by GIS Remote sensing	Rainfall monthly (liter/m2)	Volume monthly (liter
January	6089	81,0442	493478,1338
February	6089	97,1000	591241,9
March	6089	141,06	858914,34
April	6089	182 ,82	1113190,98
May	6089	164,33	1000605,37
June	6089	118,52	721668,28
July	6089	110,03	669972,67
August	6089	140,58	855991,62
September	6089	180,99	1102048,11
October	6089	222,64	1355654,96

November	6089	194 ,79	1186076,31
December	6089	128,98	785359,22
Total			9666732.894litres/year

Tableau 9   Volume of water from the ponds on Walikale Avenue.

The volume of water in these ponds is 9,666,732,894 liters of water per year with a surface area of 6089m^2 .

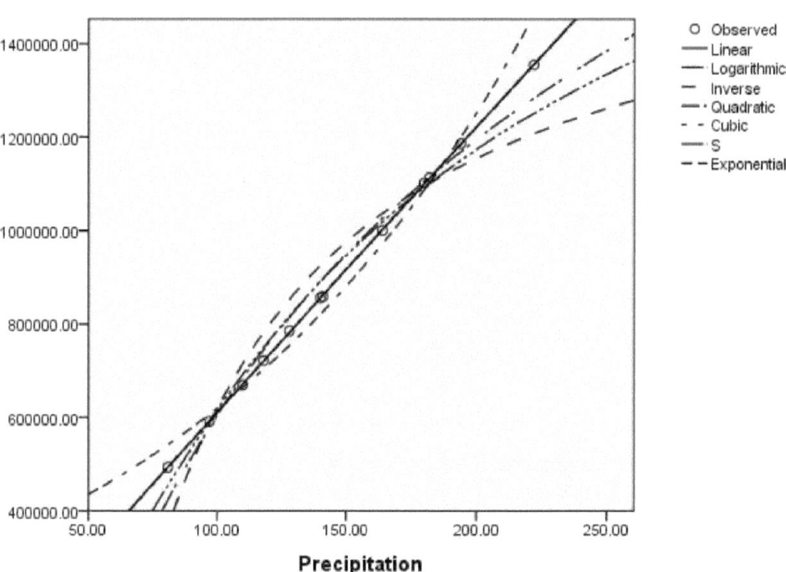

Figure 13. This graph explains the relationship between the amount of rainfall (rainfall data) and the volume of water in the Walikale Avenue water ponds (remote sensing).

4.1.2.9. The water ponds on Avenue Walikale II Commune Kisangani.

Month	Area (m^2) per GIS Remote sensing	Monthly rainfall (liter/m2)	Monthly volume (liter)
January	9256	81,0442	750145,1152
February	9256	97,1000	898757,6
March	9256	141,06	1305651,36
April	9256	182 ,82	1692181,92
May	9256	164,33	1521038,48
June	9256	118,52	1097021,12

July	9256	110,03	1018437,68
August	9256	140,58	1301208,48
September	9256	180,99	1675243,44
October	9256	222,64	2060755,84
November	9256	194 ,79	1802976,24
December	9256	128,98	1193838,88
Total			16317256,16

Tableau 10   Volume of water in the Walikale ponds II.

The volume of water in these ponds is 16,317,256.16 liters of water per year.

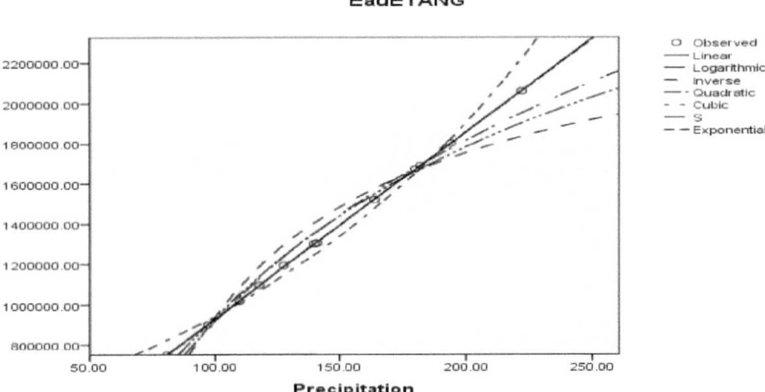

Figure 14. Relationship between the volume of water and the amount of rainfall on the Walikale Avenue II water ponds.

4. CONCLUSION AND SUGGESTION

This study carried out in the city of Kisangani precisely in the communes of Kabondo and Kisangani has its origin in the questions related to the problem of risk in flood in zones being in these communes and the months with extreme rainfalls of the city by evaluating which are the surfaces of the ponds, as well as the volume of the floodable water monthly by pond on the city of Kisangani, precisely in the communes of Kabondo and Kisangani?

Its main objective was to evaluate under the only rainy season with these four small seasons of the city of Kisangani, the surface in hectare lost in the form of water ponds in the communes of Kisangani and Kabondo; to sketch the cartography of risk of flood and water ponds in the communes of Kisangani and Kabondo as well as the level of the floodable waters We have put forward hypotheses according to which Given the planimetry of the city of Kisangani, without relief nor strong gravity; the waters of the rains (1800 to 2000 mm of water/year) stagnate in the very low curves of levelled to generate marshy and floodable zones.

For this reason, we investigate whether it is possible to calculate the space occupied by ponds in the two municipalities as our first hypothesis.

The work will also focus on verifying whether there are permanently flooded areas in Kabondo and Kisangani and calculating the volume of water in these ponds.

And these hypotheses are verified because as regards the spaces which occupy the ponds of water we had identified an agglomeration of the ponds of water in the commune of Kabondo under three places (District tshuapa, district Kibibi and 21ème trans next to the river Mukoko) and all these ponds made a surface of $41km^2$ on $368km^2$ of total surface of the commune that is to say approximately 11% of the commune ;On the commune Kisangani we had identified six agglomerations of the ponds of waters (on 13ème avenue district mosibasiba, on 10ème and 11ème avenues, on the districts Kongakonga I and II, on the avenue walikale I and walikale II) and all have a surface of 120km2 on 636km2 therefore approximately 19% of the commune. Thus, the commune of Kisangani has a larger area occupied by water ponds than the commune of Kabondo, given its position near the Congo River and its relief. This is what justifies our first hypothesis.

As for our second hypothesis, we confirm it by saying that there are floodable areas in these two municipalities by seeing the above result.

Regarding the volume of water containing these ponds we found that in the ponds of the commune Kabondo that the ponds of the district tshuapa contain more water (46.155.495,1 liters of water/year) and the ponds of the district Kibibi are the less contained of water (1246931,2 liters of water/year), on the other side in the commune Kisangani on the six agglomerations of the ponds it is the ponds of water on $10^{\text{ème}}$ and $11^{\text{ème}}$ avenues that contain more water (73.050.222,88 liters of water/year) and the ponds of water on the avenue walikale which contain less water among the six (9.666.732,894 liters of water/year).

As for the contour lines, we found that all our ponds were below 400m altitude, so we should avoid buying any land below 400m altitude because it is swampy land, difficult to build on, but rather leave it for agriculture in the broadest sense.

We suggest that this study be deepened by our scientific cadets to go through the whole city to draw more general conclusions; we then suggest to the government the following:

> increased use of renewable and clean energy,

> waste treatment,

> rehabilitation of gutters and water pipes in the cities.

> the construction of dikes and water retention basins, the

> cultivation of wood energy,

> the fight against poverty by converting ponds into fish and rice growing areas,

> wastewater treatment,

> reforestation of deforested areas,

> prevention of water-borne diseases and financing of studies on climate forecasting and modeling in at-risk areas such as Kisangani.

In the end, the planners that we are should make a warning system concerning the swampy spaces with the land registry service so as not to sell them to people in order to build houses but to leave them for agriculture in the broad sense and to feed the population.

GLOSSARY

1. GPS : Global Positioning System

2. DTM Digital Terrain Model

3. NASA : National Aeronautics and Space Administration

4. OSFAC : Central African Forest Satellite Observatory

5. SRTM : Shuttle Radar Topography Mission

6. USLE : Universal Soil Loss Equation

7. QGIS : Quantum Geographical Information System.

8. GIS : Geographic Information System.

9. LANDSAT: Earth observation satellite.

10. 3D : Three dimensions.

BIBLIOGRAPHIC REFERENCES

❖ ABDELKADER EL GAROUANI, ABDELAZIZ MERZOUK, SAID HINJE, JABRANE RAOUF, MOHAMED RACHED BOUSSEMA (1997). Identification of geohazards by remote sensing: soil erosion and landslides in the Prérif marneux (Morocco), *in* Gestion des sols dans un bassin versant. IXèmes Journées du Réseau Télédétection.

❖ KISANGANI METEOROLOGY DIVISION (2014) .Rainfall or climate data (maximum and minimum temperatures, precipitation from 1956 to 2005) for the city of Kisangani.

❖ ILUNGA, C. (2012).Remote sensing mapping of land use in the city of Kisangani and its periphery between 2002 and 2010.DEA thesis in Biodiversity Management and Sustainable Forest Management, REFORCO/ Faculty of Science.

❖ KABASELE YENGAYENGA (2009), Bas Fleuve Utilisation des Données Spatiales en Appui à la Modélisation de la Climatologie, de la Limnimétrie et de la Mare graphie en RD Congo, PhD Thesis UPN Kinshasa

❖ KABASELE, YENGAYENGA. (2013).Note de Cours de Géomantique Satellitaire, Télédétection Satellitaire et Gestion de l'environnement. University of Kisangani first degree Eaux et Forêts/ FGRNR, unpublished course.

❖ MAYOR'S OFFICE OF KISANGANI (2011). *Rapport Annuel Administration du territoire E.T.D ville de Kisangani (Exercise 2011)*. Administrative Organization (1), Kisangani, pp. 5-10.

❖ PAMANI, (2005).Identification of geo-risks by a remote sensing approach and dynamics of the erosion phenomenon on the Mont-Amba Hill.Thesis, UNIKIN, Faculty of Agronomic Sciences.

❖ SADIKI ABDELHAMID, BOUHLASSA SAÏDATI, AUAJJAR JAMAL, FALEH ALI AND MACAIRE JEAN-JACQUES (2004). Use of a GIS for the evaluation and mapping of erosion risks by the Universal Soil Loss Equation in the Eastern Rif (Morocco): case of the oued Boussouab watershed. Bulletin de l'Institut Scientifique, Rabat, section Sciences de la Terre, n°26.

WEBOGRAPHY

http://www.memoireonline.org

http://www.ese.u-psud.fr

http://www.unamur.be(science/geography/e.

http//G:/spatialModel.Learningobject.html

www.gita.info/concept Mod/en/html/....

NUMERICAL PREDICTION OF RAINFALL IN THE CITY OF KISANGANI STUDY CARRIED OUT IN THE COMMUNES OF KABONDO AND KISANGANI

1. Summary

This article is part of the adaptation to climate change, advocating a remote sensing study of the city, to identify the months with high rainfall and seek to know if there is how much season in Kisangani and measure the floodable and flooded areas by associating them with climate models of rainfall predictions to predict the risks of flooding. In addition, adaptation to climate change involves, among other things, the construction of dykes and water retention basins, the cultivation of wood energy, the fight against poverty by converting ponds into fish and rice growing areas, wastewater treatment, the prevention of waterborne diseases, and the financing of studies on climate predictions and modeling in risk areas such as Kisangani.

1. Issue

A question arises: What are the monthly numerical prediction equations between rainfall and pond water volume?

The city of Kisangani is considered as a prey to heavy floods, the government of the Democratic Republic of Congo has declared the city of Kisangani, capital of the dismembered Oriental Province and its surroundings a disaster area, after severe floods that had affected them, said Saturday, April 3, 2010 an adviser to Adolphe Muzito, former Congolese Prime Minister. According to an associate professor and environmental expert at the University of Kisangani who requested anonymity, the torrential rains that fell on Orientale province and the city of Kisangani, caused the rivers to rise to their highest historical level. Does the flooding of Kisangani have a scientific explanation? We think so, because in the era of climate change and recurrent variability of extreme weather events, Africans must fight not only to mitigate the negative effects of climate in their immediate environment, but especially to adapt to it. Thus, in the world, the mitigation of severe climate must go through reforestation, the fight against deforestation, the increased use of renewable and clean energy, waste treatment, rehabilitation of gutters and water pipes in cities.

Keywords: Geographic Information System GIS, LANDSAT satellite, Global Positioning System : GPS, Digital Terrain Model : DTM, Shuttle Radar Topography Mission : SRTM, Conceptual Data Model : CDM, United Nations Forum on Climate Change : UNFCCC ; Global Warming/Climate Change (s), Remote Sensing, Rainfall prediction, Arc Gis, USLE: Universal Soil Loss Equation, SPSS : Statistical Package for the Social Sciences.

2. Data, Software and Methodology.

1.1. Data used

1.1.1. Rainfall data for the city of Kisangani from 1956 to 2008.

1.1.2. GPS coordinates of the communes of Kisangani and Kabondo as well as those of the water ponds located in these communes.

1.2. Software

The software used was SPSS 16.0 to allow us to make statistical analyses of rainfall data (RR rainfall) to see the rainiest months and know if there are how many seasons in the city of Kisangani and the statistical software ANOVA to make numerical prediction models.

3.4. Methodology

The methodology used is the following and can be summarized in 4 main steps:

1).field data collection; 2).rainfall data processing; 3).installation

SPSS software; 4).data processing.

4.EXPECTED RESULTS

4.1.3. Presentation of monthly rainfall diagrams from 1956 to 2008.

4.1.3.1. Statistics of Monthly Precipitation Kisangani City in 52 years Source Eungi Ir2Unikis 2014

	N	Minimum	Maximum	Mean	Std. Deviation
JanRains	52	13.80	182.60	81.0442	45.54183
Rainfall	52	20.80	196.30	97.1000	42.15289
MarchRains	52	46.30	299.30	1.4106E2	56.10866
AprilRains	52	60.50	315.10	1.8282E2	58.40811
MayRains	52	81.10	311.00	1.6433E2	53.06103
JunePews	52	39.60	302.00	1.1852E2	53.60688
JullPluies	52	46.70	234.00	1.1003E2	38.82213
AugustPeople	52	44.40	286.00	1.4058E2	48.80039
SeptPluies	52	80.30	337.50	1.8099E2	58.63398
OctPluies	52	94.50	375.50	2.2264E2	59.35511
NovRains	52	34.20	388.30	1.9479E2	66.86915
DecRains	52	29.40	346.20	1.2898E2	65.42602
Valid N (listwise)	52				

Table 1: Descriptive statistics of monthly prescriptions in the city of Kisangani in 52 years.

Comment:

1.   N = 52 years of rainfall data from Kisangani 1956 -2008

2.   Minimum, Maximum and Mean in mmH2O = $LitreH20/m^2$ from January to December over 52 years

3.   Jan =January, Feb = February, Jull= July, Sept = September, Oct = October, Nov= November, Dec= December and, Valid N Listwise = Total validated N

4.   Std.Deviation = Standard Deviation

5.   Monthly rainfall = Mean ± Standard deviation

6.   Minimum = Lowest monthly value for 52 years

7.   Maximum= Highest monthly value for 52 years

8.   Mean = 1.4106E2 = 1.4106x100 =141.06 for example

4.2.3.2.

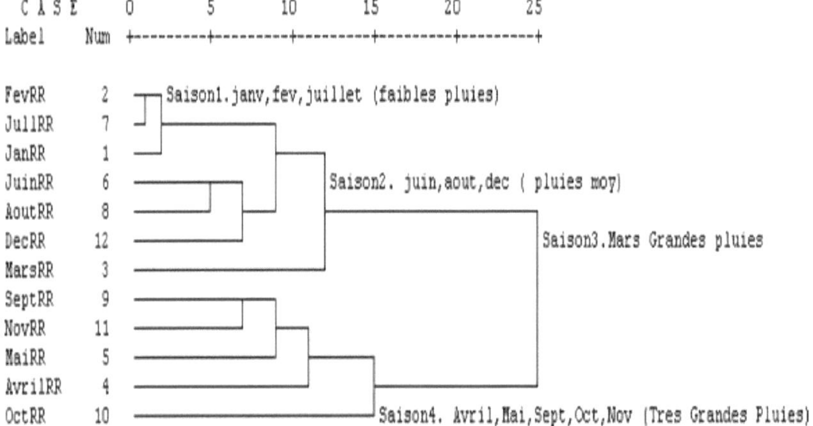

Figure 18.Euclidean distance of rainfall in Kisangani

Comment:

1.   The rainfall of February and July are close on a small scale of 1/25 while January and February are close on the order of 2/25

2.   Kisangani is characterized by a single annual rainy season subdivided into four groups of

short seasons: season 1 (January, February, July); season 2 (June, August, December); season 3 (March); season 4 (April, May, September, October, November)

3.   March, June, and October are difficult months to predict from the others (see ANOVA model below)

4.1.3.3 **Component Plot Source Eungi Ir2 Unikis2014**

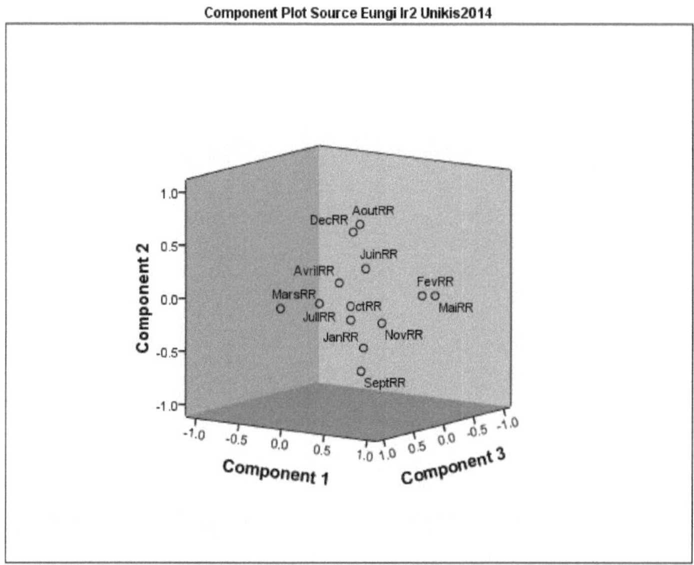

Figure 19.Component plot

Comment:

1.   Principal Component Analysis = Spatial Representation of the Dendrogram, i.e. Euclidean distances between the rainy months

2.   In spite of the month of March, which is different from the modeling, the rainfall in Kisangani, overlaps on three main axes (Compnent 1,2,3)

4.1.3.4 Numerical rainfall forecasts in Kisangani

Calculated by ANOVA regression equations, the probable rainfall amounts to be collected in the next months knowing the in situ values of the closest recent months according to the following graphical correlations:

0. January and December

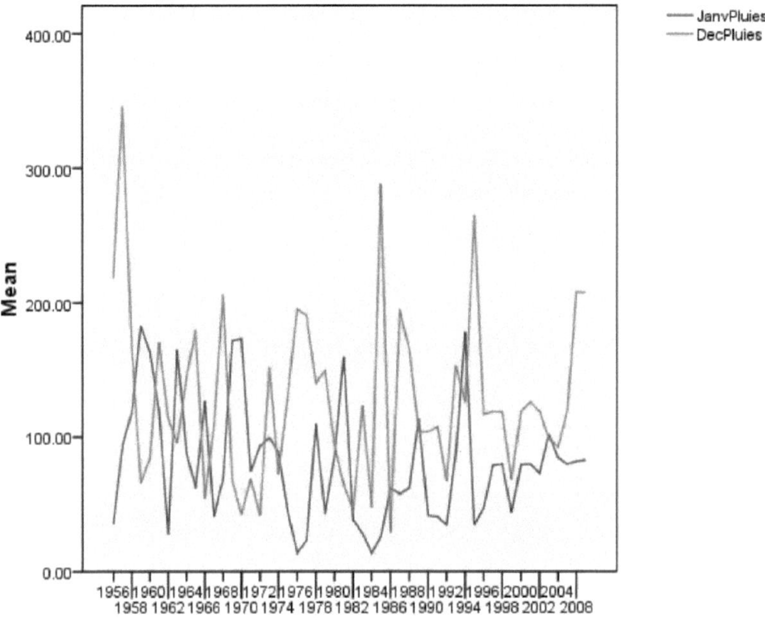

Figure 20: Kisangani Monthly rainfall for January and December.

This figure shows us that December has a fairly high rainfall than January and these two months are closer in recent times.

4.1.3.5. Point cloud of correlation between January and December, cubic model.

JanRains

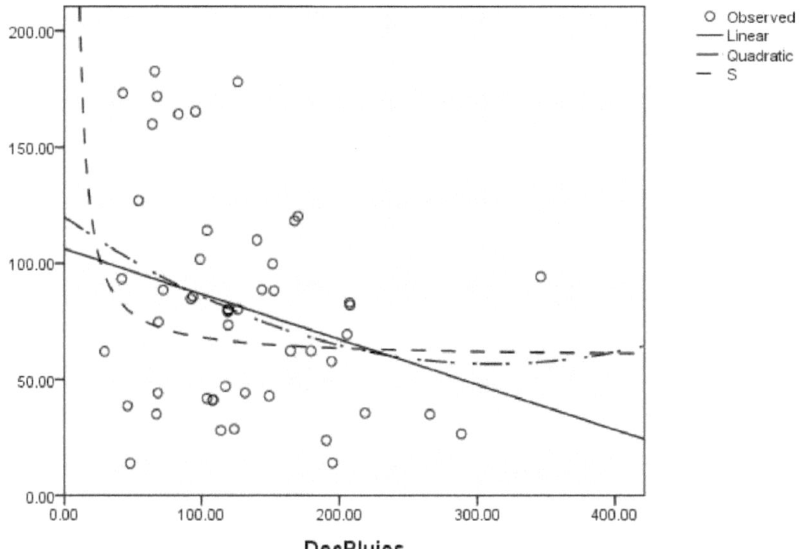

Figure 21.Correlation point cloud between January and December, cubic model.

The figure above explains the point cloud which is the rainfall in January and December.

Table 2.The ANOVA model under SPSS 16.0 of the numerical rainfall forecasts in January.

Model		Unstandardized Coefficients		Sia.
		B	Std. Error	
1	(Constant)	106.152	13.642	.000
	DecRains	-.195	.095	.045

a. Dependent Variable: JanRain

This table shows the December rainfall statistic as a dependent variable of January which correlates with the figure above.

Our numerical predictions of rainfall in January, based on the ANOVA model under SPSS 16.0 :

JanRain = 106.152 - 0.195xDecRain, in mm of water

Table 3.Numerical rainfall forecast in December

Model		Unstandardized Coefficients		Sia.
		B	Std. Error	
1	(Constant)	161.541	18.092	.000

Januaries	-.402	.195	.045

a. Dependent Variable: DecPluies

Our numerical predictions of rainfall in December, based on the ANOVA model under SPSS

16.0 :

DecRain = 161.541 - 0.402xYearRain, in mm of water ;

4.1.3.6. Superposition of months with similar monthly rainfall.

2. February and May

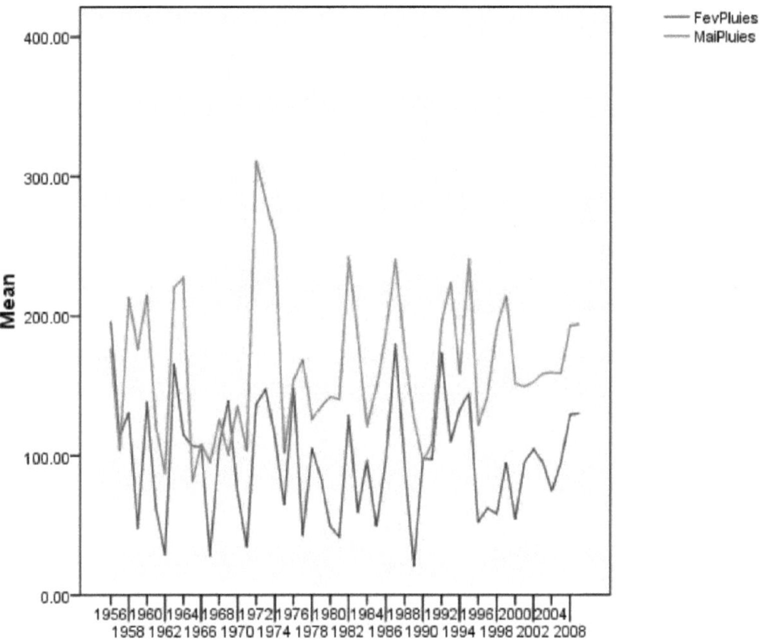

Figure 22: Illustrates the monthly rainfall in superposition of the months of February and May. It follows from this figure that the months of February and May are close in rainfall and superimposable.

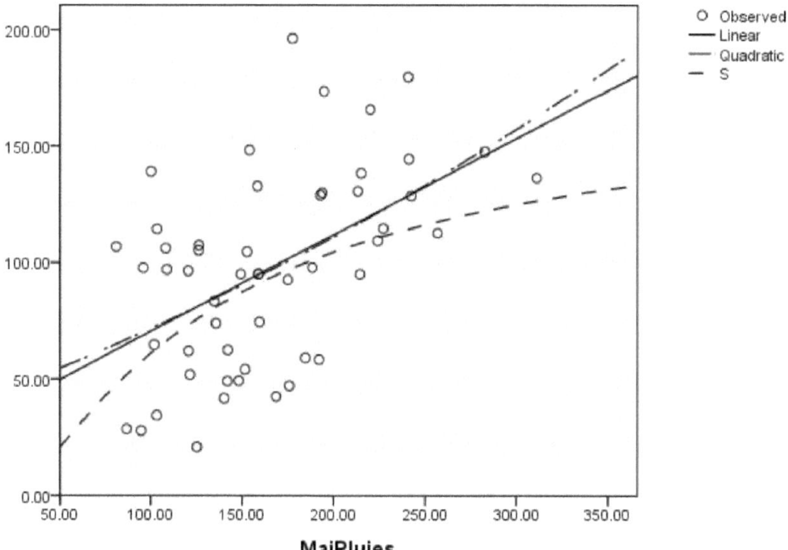

Figure 23. Cloud of correlation points between February and May. The month of May has a higher rainfall than February.

	Unstandardized Coefficients		
Model	B	Std. Error	Siq.
1 (Constant)	29.116	16.547	.085
MayRains	.414	.096	.000

a. Dependent variable: FevPluies

Table 4: Numerical rainfall forecast for February

Our numerical predictions of rainfall in February, based on the ANOVA model under SPSS 16.0 :

Rainfall = 29.116 + 0.414xMayRainfall, in mm of water ;

3. February, May and March

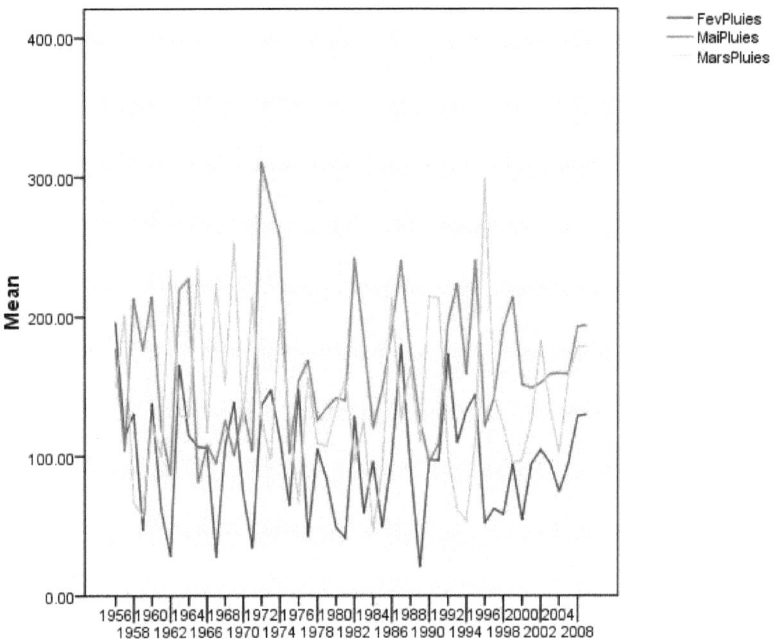

Figure 24: Illustrates the monthly precipitation of the months of February, March and May. It follows from this figure that the month of May has a large monthly precipitation compared to the others and the three months are superimposable near recent.

Table 5: Numerical rainfall forecast for February

Model		Unstandardized Coefficients		Sia.
		B	Std. Error	
1	(Constant)	100.677	16.063	.000
	FevPluies	.656	.152	.000
2	(Constant)	147.752	23.418	.000
	FevPluies	.591	.146	.000
	MarchRains	-.289	.109	.011

a. Dependent Variable: MayRain

Our numerical predictions of rainfall in February, based on the ANOVA model under SPSS

16.0 :

According to the model1 : May Rainfall = 100.677 + 0.656xFebRainfall, in mm of water ;

According to the model2 : May Rainfall = 147.752 + 0.5916xFebRainfall - 0.289xMarchRainfall, in mm of water ;

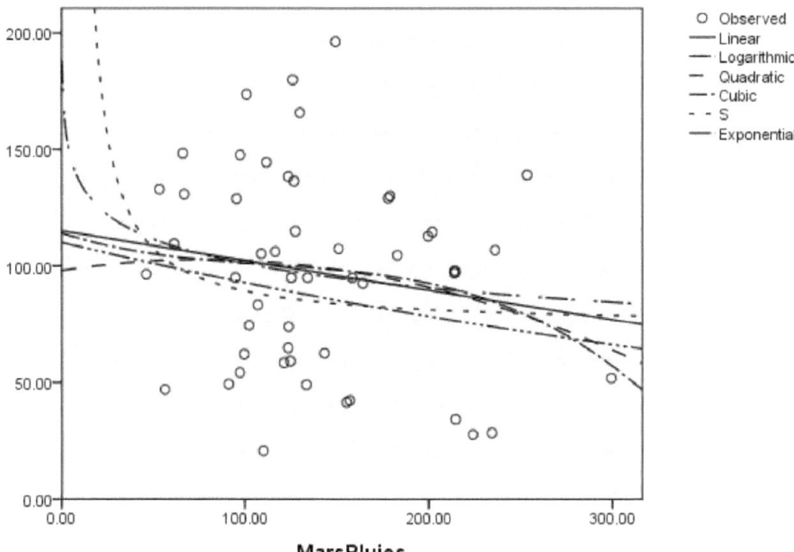

Figure 25: Illustrates the correlation point cloud between February and March.

It is not easy to compare the rainy month of March with the month of February.

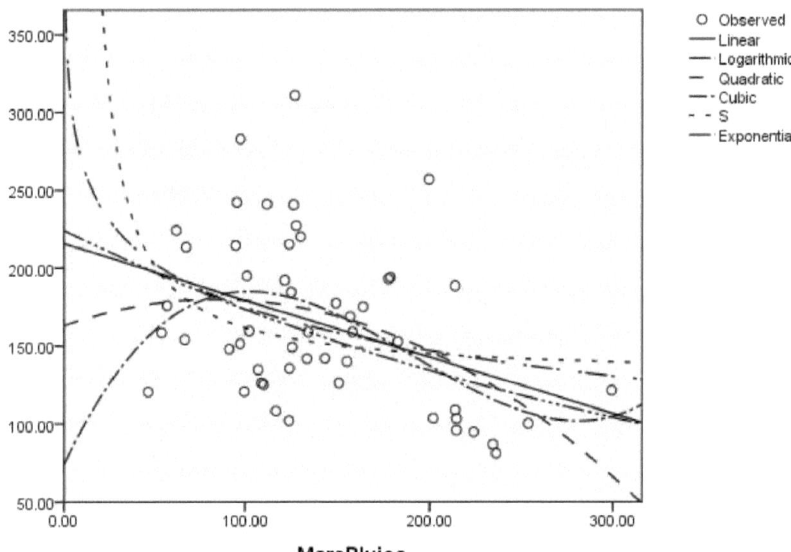

Figure 26. Illustrates the correlation point cloud between the months of May and March, cubic model.

It is easier to compare the rainy month of March with the previous month of May, according to the logarithmic ANOVA model below:

Table 6. Numerical rainfall forecast in May, logarithmetic model.

Coefficients

	Unstandardized Coefficients		
	B	Std. Error	Sig.
MarchRains	-.003	.001	.001
(Constant)	224.170	24.710	.000

The dependent variable is ln(MaiPluies) = ?

Log (May Rains) = 224.17 - 0.03 x March Rains.

In addition, the multi-variate linear ANOVA model of March versus recent May, April and July is the best according to the graphs and equations:

4. April, May, July.

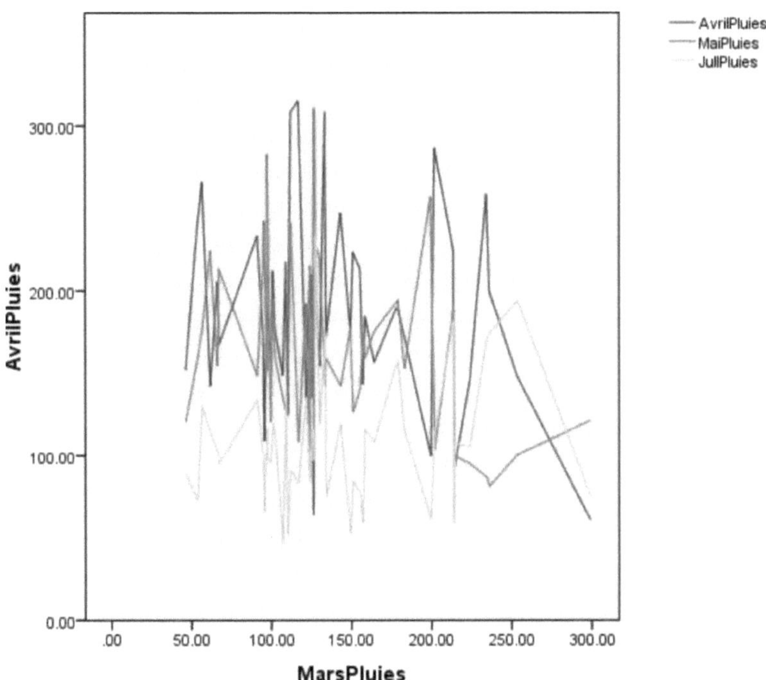

Figure 27: Illustrates the monthly precipitation of the months of April, May and July. These months are really close recently just by seeing their overlap.

Table 7: Numerical rainfall forecast for March

Coefficients[a]

Model	Unstandardized Coefficients		Sig.
	B	Std. Error	
1 (Constant)	207.925	23.814	.000
MayRains	-.407	.138	.005
2 (Constant)	259.631	32.997	.000
MayRains	-.427	.133	.002
AprilRains	-.265	.121	.034
3 (Constant)	225.836	33.183	.000
May Rains	-.437	.125	.001

| April Rainfall | -.375 | .120 | .003 |
| Jull Rains | .505 | .180 | .007 |

a. Dependent Variable: March Rainfall

Modele1:MarsPluies=207,9250,407xMaiPlui es

Model2:March Rain=259.631 -

0.427xMay Rain-0.265x April Rain.

Model3: March Rainfall = 225.836 +0.505xJuly - 0.437xMay - 0.375xApril

Table 8: Numerical rainfall forecasts for April.

Coefficients[a]

Model	Unstandardized Coefficients		Sig.
	B	Std. Error	
1 (Constant)	128.516	23.426	.000
JullPluies	.494	.201	.018
2 (Constant)	165.568	26.415	.000
JullPluies	.605	.195	.003
March Rains	-.350	.135	.013

a. Dependent Variable: April Rainfall

Model1:MarchRains=128.516+0.494xJulyRains

Model2:MarsRains=165.568+0.605xJulyP luies-0.350xMarsRains

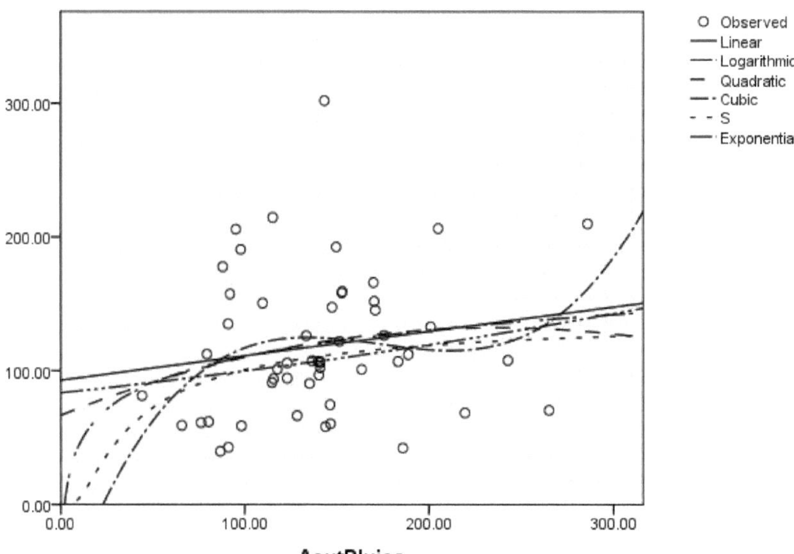

Figure 28.Illustrates the correlation point cloud between June and August, cubic model The projection of the months on the above axes is superimposable.

Table 9: Numerical rainfall forecasts for June.

Coefficients

	Unstandardized Coefficients		Sig.
	B	Std. Error	
1 AugustRains	/-33.621	19.321	.088
(Constant)	4.947	.168	.000

The dependent variable is ln(June Rainfall).

Model1:Log(JuneRain)=4.94733.621xAugustRain

5. November, March and April

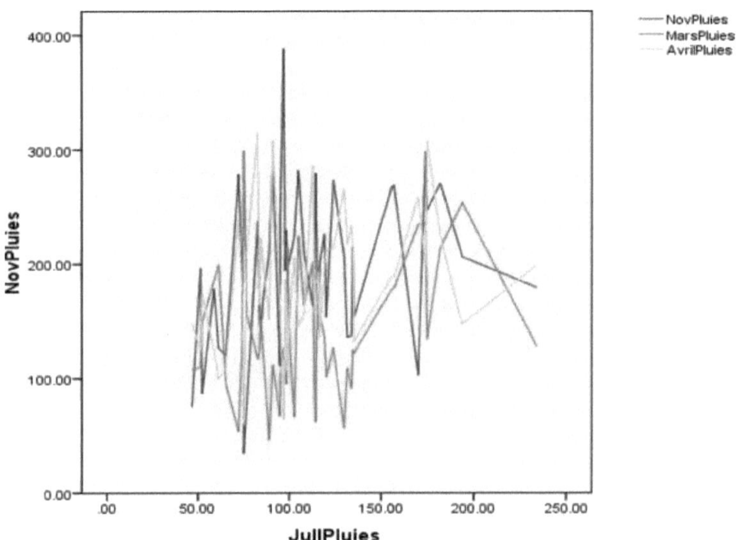

Figure 29.*Illustrates the monthly precipitation for the months of March, April and November.*

Table 10. Numerical rainfall forecast in July.

Coefficients[a]

Model	Unstandardized Coefficients		Sig.
	B	Std. Error	
1 (Constant)	70.168	17.027	.000
April Rains	.218	.089	.018
2 (Constant)	29.184	23.354	.217
April Rains	.271	.087	.003
March Rains	.222	.091	.018
3 (Constant)	-5.082	26.290	.848
April Rains	.253	.084	.004
March Rains	.246	.087	.007
Nov Rainfall	.175	.071	.018

Coefficients[a]

42

	Unstandardized Coefficients		Sig.
Model	B	Std. Error	Sig.
1 (Constant)	70.168	17.027	.000
April Rainfall	.218	.089	.018
2 (Constant)	29.184	23.354	.217
April Rains	.271	.087	.003
March Rains	.222	.091	.018
3 (Constant)	-5.082	26.290	.848
April Rains	.253	.084	.004
March Rains	.246	.087	.007
Nov Rainfall	.175	.071	.018

a. Dependent Variable: Jull Rainfall

Model1:JulyRains=70,168

+0.218xApril rains

Model2:JulyRain=29.184+0.271xAprilRain

ies+0.222xMarchRains

Model3:JulyRains= -5.082

+0.253xAprilRains

+0,246xMarsPluies+0,175xNovembre

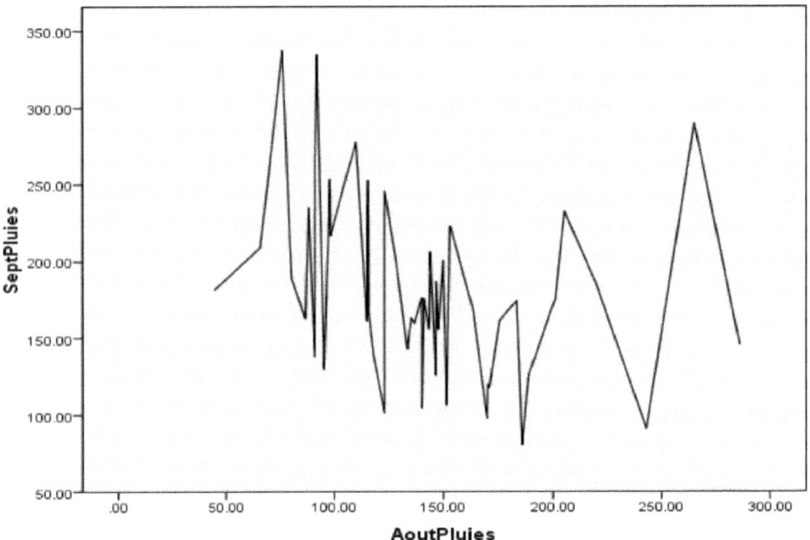

Figure 30.Monthly precipitation between August and September.

The two months are almost the same.

Table 11.Numerical rainfall forecast in August.

Coefficients[a]

Model	Unstandardized Coefficients		Sig.
	B	Std. Error	
1 (Constant)	185.015	21.377	.000
Seven Rainfalls	-.246	.112	.034

a. Dependent Variable: AugustRains Model1:AugustRains=185.015 - 0.246xSeptemberRains

6. August and **November**

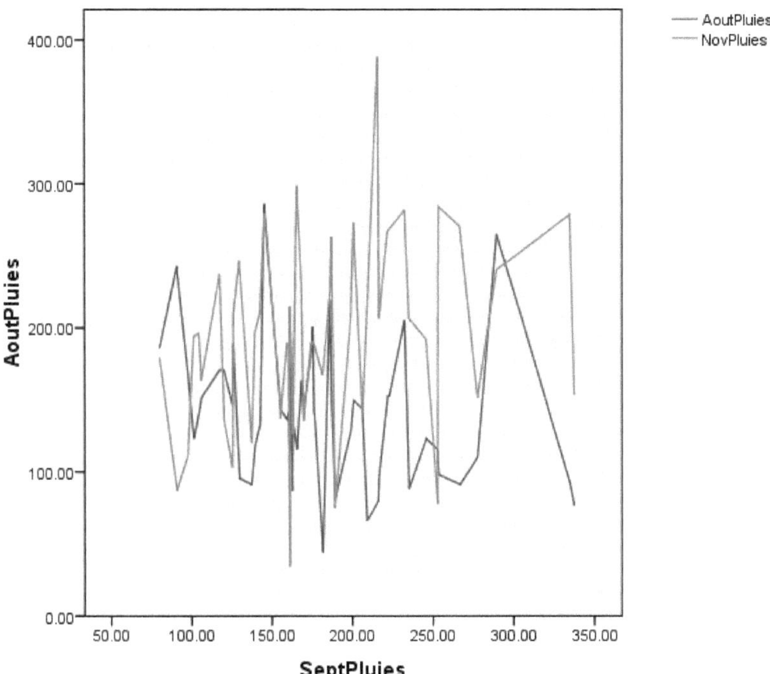

Figure 31.Monthly precipitation for August and November. The two months are close to each other and superimposable, the month of November has a higher rainfall than the month of August.

Table 12: Numerical rainfall forecast for September

Coefficients[a]

Model	Unstandardized Coefficients		Sig.
	B	Std. Error	
1 (Constant)	230.817	24.134	.000
AugustRains	-.354	.162	.034
2 (Constant)	180.409	31.014	.000
AugustRains	-.377	.155	.019
NovRains	.275	.113	.019

a. Dependent Variable: RainySeven Model1:RainySeven=230.817- 0.354xAugustRainy

Model2:SeptRain=180.409 +0.275 x NovRain -0.377x AugRain

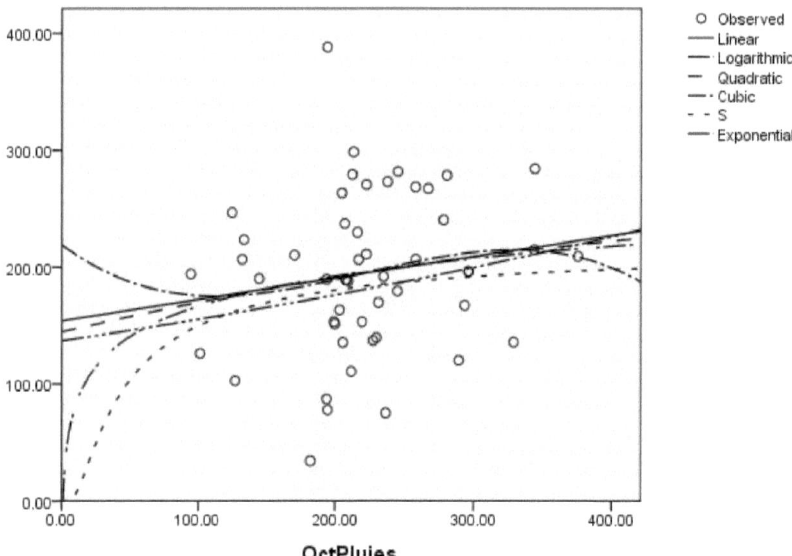

Figure 32.Illustrates the cloud of correlation points between November and October.

From this graph, it can be seen that the points are clouds of points that represent the rainfall observed during the two months mentioned above.

Table 13: Numerical rainfall forecasts for November

Coefficients

	Unstandardized Coefficients		
	B	Std. Error	Sig.
OctRains	.001	.001	.209
(Constant)	137.072	31.083	.000

The dependent variable is ln(NovPluies).

Model1:NovRain=137.072 + 0.001 OctRain

7. July and September

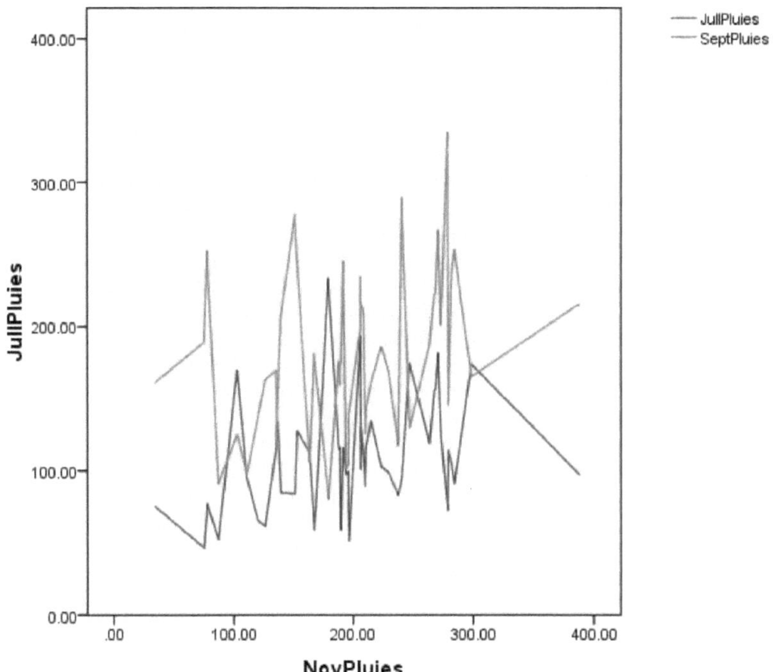

Figure 33.Monthly precipitation for the months of July and September. These months are superimposable and the month of September has a higher rainfall than the month of July.

Table 14: Numerical rainfall forecasts for November

Coefficients[a]

Model	Unstandardized Coefficients		Sig.
	B	Std. Error	
1 (Constant)	138.679	27.118	.000
JullPluies	.510	.233	.033
2 (Constant)	69.733	38.285	.075
JullPluies	.545	.222	.018
Seven Rainfalls	.360	.147	.018

a. Dependent Variable: NovRain

Model1:NovRains=138.679 +0.510 x

JulyRains

Model2: NovRain=69.733 +0.545

xJulyRain + 0.360 SeptRain

Std.Error is the error deviation of the calculated coefficient.

3. CONCLUSION AND SUGGESTIONS

This study carried out in the city of Kisangani precisely in the communes of Kabondo and Kisangani originates from questions related to the problem of flooding risk in areas located in these communes and the months of extreme rainfall in the city by evaluating What are the monthly numerical forecasting equations between the rainfall and the volume of water in ponds?

We also calculated the monthly numerical forecast to predict the amount of rain that will fall in the same month next year by knowing the amount of rain that fell in that month the previous year. And from the monthly numerical equations we combined the results of the remote sensing and the rainfall data to see the relationship between them.

Among the four small seasons that the only rainy season of the city of Kisangani identified, we notice that it is the 4ème small season that is the rainiest that includes the following months:

April, May, September, October, November.

We suggest that this study be deepened by our scientific cadets to go through the whole city; we then suggest to the government the following:

> increased use of renewable and clean energy,

> waste treatment,

> rehabilitation of gutters and water pipes in the cities.

> construction of dikes and water retention basins,

> the cultivation of wood energy,

> wastewater treatment,

> reforestation of deforested areas,

> prevention of waterborne diseases and funding of forecasting studies in at-risk areas such as Kisangani.

GLOSSARY

1. GPS : Global Positioning System

2. SRTM : Shuttle Radar Topography Mission

3. SPSS : Statistical Package for the Social Sciences

4. GIS : Geographic Information System.

BIBLIOGRAPHIC REFERENCES

❖ Division de la Météorologie de Kisangani(2014). Rainfall or climate data (maximum and minimum temperatures, precipitation from 1956 to 2005) of the city of Kisangani.

❖ Mairie de Kisangani (2009). *Process of review and community restitution of the PDU Kisangani - Plan - Programme d'Action Prioritaire pour le Développement Urbain de Kisangani à l'horizon 2012*. Technical and Financial Support to the PAIDECO TSHOPO Project, Kisangani, 46 p.

❖ Kabasele YY (2009), Bas Fleuve Utilisatisation des Données Spatiales en Appui à la Modélisation de la Climatologie, de la Limnimétrie et de la Mare graphie en RD Congo, PhD Thesis UPN Kinshasa

Printed by Books on Demand GmbH, Norderstedt / Germany